DIAMON

by Ellen Ungaro

Table of Contents

Introduction

No one knows for sure when the first diamonds were discovered. Stories about diamonds started to spring up in India more than 2,000 years ago. It is easy to tell from those early tales that people thought diamonds were special. One myth says that diamond seeds grew from the bones of a great king.

For a long time, diamonds were so rare that only royal families owned them. Kings and queens wore diamond crowns and fancy jewelry.

◀ The British Queen's crown features an enormous diamond called the Koh-i-Noor.

People still prize diamonds today. Many people think diamonds are the world's most beautiful gems. Diamond rings are seen as a symbol of love. People pay thousands of dollars for diamond jewelry.

What makes diamonds special? Diamonds are made by amazing forces inside Earth. As a result of their formation, they have special **properties**, or qualities, that make them unique, or unlike any other material.

Diamonds have a long history. People have been mining them since ancient times. Today, people are still trying to find new diamonds.

▲ The Hope Diamond is one of the most famous diamonds in the world. Thousands of people come to see it at the Smithsonian Museum in Washington, D.C.

Diamonds are used for more than jewelry. Some of the ways people use diamonds may surprise you.

When you learn the story of the diamond, you will understand why it is a remarkable rock.

A Remarkable Rock

Diamonds are **minerals**. There are about 4,000 known minerals. Minerals form all the rocks on Earth. Gold, silver, and iron are examples of minerals.

Minerals have characteristic properties. A property is anything that can be measured or observed. What color is a mineral? How heavy is it? How strong is it? Does it bend, or is it brittle? Is it shiny or dull? These are all properties.

MOHS Hardness Scale

This scale is used to determine how hard a mineral is. A mineral's hardness is tested by seeing if it can scratch another mineral. For example, a mineral ranked 7 on the scale can scratch feldspar but not topaz.

1. Talc
2. Gypsum
3. Calcite
4. Fluorite
5. Apatite
6. Feldspar
7. Quartz
8. Topaz
9. Corundum
10. Diamond

▲ Many people think of diamonds as clear or colorless, but diamonds are also yellow, pink, brown, and blue.

▲ Scientists sort minerals into categories based on their luster.

One of the most important properties of diamonds is that they are the hardest known material. A diamond can scratch all other materials. Only a diamond can cut a diamond.

Diamonds are used for jewelry because of the way they sparkle and shine. People talk about the "brilliance" and "fire" of a diamond.

A diamond is said to be brilliant because of the amazing way it reflects light. This property is called **luster**. Some minerals, such as chalk, have a dull luster. Other materials shine and reflect light. But no other mineral reflects as much light as a diamond does.

▲ People say diamonds have "fire" because of the colors they see in them.

Gemstones

Diamonds are also called gemstones, or gems. A gemstone is any rare stone that is used for jewelry. About seventy minerals are considered gems. The rarest gems are called precious stones. Other gems are called semiprecious.

Here are examples:

Precious	Semiprecious
Diamond	Opal
Emerald	Garnet
Ruby	Lapis lazuli
Sapphire	Turquoise

Look closely at a clear diamond and you will see a rainbow of colors. You can see the colors because of the way diamonds **refract** (rih-FRAKT), or bend, light. When light enters a diamond, the light bends and is separated into red, orange, yellow, green, blue, indigo, and violet. Very few materials refract light the way that diamonds do.

✓ POINT

Visualize

Close your eyes and try to picture another place you've seen a rainbow. In a puddle of water? On a drinking glass? Near a window? What might this have in common with a diamond?

Inside a Diamond

All minerals are made up of one or more **elements** (EH-leh-ments). Diamonds are made up of the element carbon, which is one of the most common elements on Earth. Graphite, which is the lead in your pencil, is also made up of pure carbon.

So how can graphite, one of the softest minerals, and diamond, the hardest, be made up of the same element and have such different properties? The answer is their crystal structure.

All minerals have a crystal structure. In a crystal, the **atoms** of the element are arranged in a repeating pattern. An atom is the smallest part of an element. Diamonds have a cubic crystal system. That structure gives them their strength.

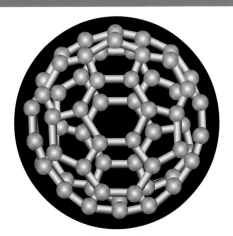

diamond

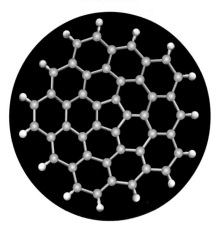

graphite

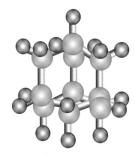

carbon atoms

▲ The pattern of carbon atoms in diamonds is different from the pattern of carbon atoms in graphite.

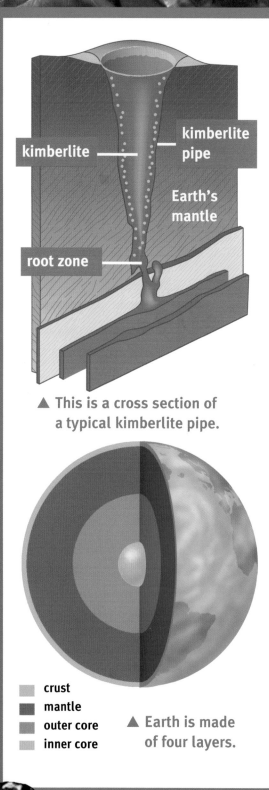

▲ This is a cross section of a typical kimberlite pipe.

kimberlite

kimberlite pipe

Earth's mantle

root zone

crust
mantle
outer core
inner core

▲ Earth is made of four layers.

How Do Diamonds Form?

What turns carbon into diamonds? It takes intense heat and pressure. Scientists think that most diamonds are one to three billion years old. They formed deep below ground in the layer of Earth called the **mantle**. It is very hot there, and the pressure is very high. That heat and pressure turned carbon into diamonds.

Diamonds were carried to Earth's **crust** in volcanic pipes. These thin pipes run from the mantle to Earth's crust. Molten, or liquid, rock erupted through the pipes at high speeds.

Most diamonds were carried in **kimberlite** (KIM-ber-lite) **pipes**. The molten rock inside is called kimberlite.

Diamonds didn't form in the kimberlite pipes. The pipes simply carried the diamonds to the surface. In fact, while diamonds are more than one billion years old, most of the pipes formed about 100 million years ago.

For diamonds to make it safely to the surface, conditions had to be right. If the liquid rock were too hot, the diamonds would burn up. If the rock moved too slowly, the diamonds would turn to graphite.

▼ Kimberlite rock is sometimes called "blue ground" because of its color.

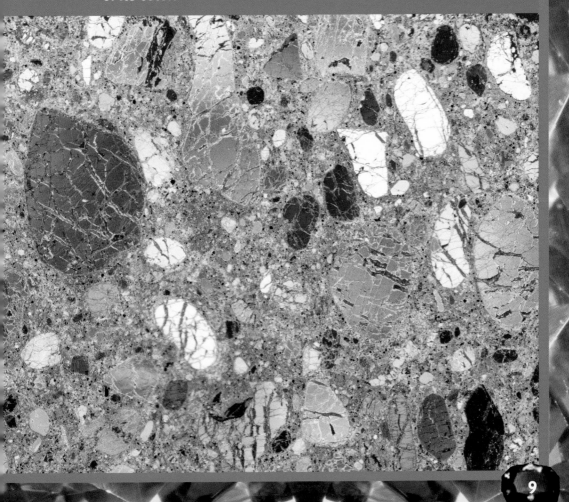

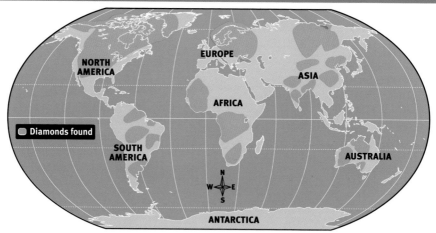

▲ Diamond pipes are found in parts of Earth's crust.
The rock there is more than two billion years old.

Where Diamonds Are Found

Most diamonds are still trapped underground in kimberlite pipes. Pipes are found in groups. There might be anywhere from six to forty pipes in one place. The pipes, which are cone-shaped, come in different sizes. Some are only two acres wide at the top. That is about the size of two football fields. Others might be thirty acres wide. Pipes stretch deep into the ground. Many are more than 3,500 feet (1,067 meters) deep.

EVERYDAY SCIENCE

Scientists believe there are diamonds in space. One reason is because diamonds have been found in craters created by meteorites that have crashed into Earth. Scientists think diamonds can be found on the moon and on the planets Uranus and Neptune.

Some diamonds ended up in rivers and streams and even the ocean. How did that happen? Wind and water **eroded** the kimberlite rock. Bits of rock—with diamonds trapped inside—broke off. Some of the rock was washed into rivers and carried away by the water. Over time, the rock wore away, and the diamonds were left. Diamonds found in kimberlite rock are called **primary deposits**.

Some diamonds were carried all the way to the ocean. Other diamonds sank to the bottom of rivers and streams and were buried in mud. When rivers changed course or dried up, hidden gems were left behind. These deposits are found far from the kimberlite pipes. They are called **secondary deposits**.

▲ It took thousands of years for the kimberlite surrounding these diamonds to erode.

Discovering Diamonds

▲ After diamonds were discovered there, Brazil quickly replaced India as the source for most diamonds.

▲ In the 1600s, an explorer named Jean-Baptiste Tavernier brought diamonds back from India to Europe.

For a long time—almost 2,000 years—all of the world's diamonds came from India. These diamonds came from secondary deposits. They were found in rivers and streams.

India had many big diamond mines. One mine had more than 60,000 people working there. Workers mined for diamonds by sifting through the mud in a riverbed. They did all the work by hand.

In the 1700s, diamonds were discovered in Brazil. Those diamonds were also found in rivers.

Then, in 1867, a fifteen-year-old boy named Erasmus Jacobs found a diamond by the Orange River on his father's land in South Africa. That discovery would change everything.

Diamond Rush

Erasmus Jacobs' discovery started a diamond rush. The next year, people found more diamonds by the Orange River. Soon, thousands of people rushed to South Africa looking for diamonds.

As more and more people came, they searched farther and farther from the river. They also began to dig deeper into the ground. Soon, people discovered the true source of diamonds. Until that time, no one really understood where diamonds came from.

When people dug deeper, they reached kimberlite rock. They saw that the diamonds were in the rock. People soon figured out that diamonds came from deep inside Earth and were carried to the surface in kimberlite pipes.

◀ By 1872, more than 50,000 people were in South Africa looking for diamonds.

☑ POINT

Reread
Why does the author say, "That discovery would change everything"? Is this statement a fact or the author's opinion? How do you know?

▲ The Kimberley mine was the first diamond mine in Africa. It operated from 1871 to 1914. It is the largest man-made hole in the world. The mine is 1,510 feet (460 meters) wide and 3,510 feet (1,070 meters) deep!

Modern Mining Begins

Finding the pipes, the true source of the diamonds, meant finding many more diamonds. The mines in India produced between 50,000 and 100,000 **carats** (KAIR-uts) a year. A carat is a unit of weight used for diamonds and other gems. By 1900, the mines in South Africa were producing two to three million carats a year.

The discovery of the pipes changed how diamonds were mined. Miners dug rock out of the ground. Then they removed diamonds from the rock. It was harder than mining diamonds from a river.

After the big find in South Africa, people also started to look for and find diamonds in other places in the world.

HISTORICAL PERSPECTIVE

It took a long time for diamonds to make their way from India to Europe. When the first diamonds started to appear, people were not sure what to make of them. Some thought they were a symbol of strength. Others believed that diamonds had healing powers. They thought placing a diamond in the mouth of a liar could cure him. Still others thought diamonds had evil powers and were poisonous.

Diamonds were still so rare, though, that kings and queens owned most of them. That finally changed when diamonds were found in South Africa.

Eyewitness Account

In 1851, the Koh-i-Noor diamond was put on display at the World's Fair in England. People flocked to see the famous diamond. It was the first time many people saw a real diamond. Many were disappointed. One reporter wrote, "After all the work which has been made about this celebrated diamond, our readers will be rather surprised to hear that many people find a difficulty in bringing themselves to believe . . . it is anything but a piece of common glass."

In the Diamond Mines

Today, there are diamond mines in many countries. Almost all diamonds are mined from pipes within Earth. Most diamond mines are open pit mines. That means miners get to the diamonds by digging a large hole.

Once a pipe with diamonds has been found, workers begin digging. First, they remove the loose soil above the pipe. Then they start to cut out huge chunks of rock. Sometimes they use explosives to break up the rock. Large trucks carry the loose rock out of the mine. The next step is to find the diamonds in all that rock.

Diamond Mining Today
These five countries produce most of the world's diamonds.

Country

Country	% of World's Diamonds
Australia	~28
Botswana	~24
Russia	~14
Congo	~13
South Africa	~9

% of World's Diamonds
(scale: 5, 10, 15, 20, 25, 30)

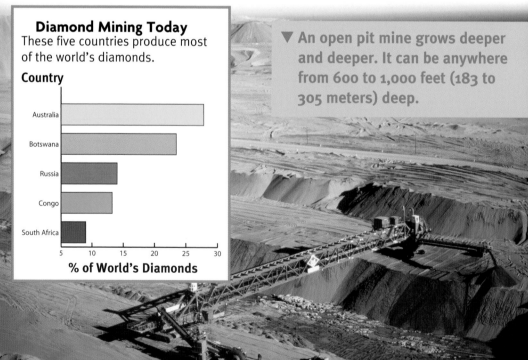

▼ An open pit mine grows deeper and deeper. It can be anywhere from 600 to 1,000 feet (183 to 305 meters) deep.

How Diamonds Are Mined

①

▲ Large pieces of rock are cut or blasted out of the ground.

②

▲ The rock is crushed into smaller pieces.

③

▲ The crushed rock is mixed with a liquid and placed in a machine that spins it around at high speeds. This separates the materials.

④

▲ The crushed rock is placed on a grease table, which is a large moving belt covered with grease. The table is sprayed with water. The rocks wash off, but the diamonds stick to the grease.

▲ In southern Africa, many diamonds were carried to the sea by the Orange River in South Africa.

Diving for Diamonds

After years of mining, most of the diamond deposits in rivers were gone. But people knew that some of the diamonds picked up by rivers ended up in the ocean.

In the 1960s, people started to search for diamonds along the coast of southern Africa. Before that, people had picked up single diamonds on the beach there. A few fishermen had scooped up gems in shallow water. When people began to look more closely, they found big deposits of diamonds on the beach and in the ocean. People have been mining the waters there ever since.

In shallow waters, scuba divers find the diamonds. They dive down carrying large vacuum hoses. The hoses suck up sand, small rocks, and—hopefully—diamonds from the ocean floor. Special mining ships are used in deeper waters.

Mining diamonds from the ocean can be difficult. But it is worth the effort because the diamonds found in the ocean can be especially beautiful. Of the diamonds found on the ocean floor, ninety-five percent are gem quality.

Diamonds in America

The United States isn't home to any diamond mines. But there is one place where you can dig for them yourself. Crater of Diamonds State Park in Arkansas is the only public diamond mine in the world where anyone can hunt for diamonds. And people do find diamonds there. Most of them are smaller than a pencil eraser. But a few lucky searchers have found diamonds as big as six carats.

▼ Mining ships carry drills that dig into the ocean floor and pumps that bring the rocks and sand to the surface.

OCEAN STROOM
CAPE TOWN

19

Searching for New Diamonds

Mining companies are always searching for new diamond pipes. Most mining companies hire geologists (jee-AH-lih-jists) to help them.

When hunting for diamond pipes, the first thing geologists look for are indicator minerals. These are the minerals that are usually found with diamonds. Garnet, a bright red stone, is an indicator mineral for diamonds. When geologists find garnets, it is a sign that there may be diamonds close by.

Indicator minerals can help a geologist find a pipe. But sometimes

▲ Garnet is an indicator mineral for diamonds, but it is also a popular gemstone.

the minerals have been carried far from their source. Geologists have to understand how the land has changed over thousands of years. Knowing where rivers once ran can help them trace a mineral back to its source.

IT'S A FACT

Termites can help geologists find diamonds. When termites are building their homes, they bring soil up from deep underground. Geologists take samples of this soil to look for indicator minerals.

Another way geologists find diamond pipes is to survey the land from a plane. This lets them view large stretches of land. One thing they look for are round, bowl-shaped holes in the ground that might be pipes.

Once geologists have found a kimberlite pipe, they take samples to see if the pipe has diamonds. There are only about 6,000 kimberlite pipes in the world. Most have no diamonds.

If geologists do find diamonds, they dig deeper and take more samples. They want to find out how many diamonds are there. If there are too few, starting a mine will not be worth it. If the sample looks good, they will build a mine. The whole process takes about ten years.

They Made a DIFFERENCE

Geologist John Gurney made a discovery that changed the way people looked for diamonds. Scientists had known for a long time that garnet was an indicator mineral for diamonds. In 1973, Gurney decided to take a closer look at garnets. He studied the chemical makeup of the gem. Gurney discovered that the garnets found with diamonds are different from other garnets. Geologists call these garnets G-10. Now when geologists find G-10 garnets, they know there are diamonds nearby.

Geologists use special ▶ equipment to survey the land.

From Mine to Store

What happens to diamonds after they are mined? First, diamonds are sorted into two categories—gems and industrial diamonds. Gems are the diamonds that will be used for jewelry. Only fifteen to twenty percent of diamonds are gems. The rest are industrial diamonds. They will be used for tools.

Gem diamonds are then sorted into categories. They are sorted by weight. Diamond weights are measured in carats. The higher the number of carats, the more a diamond is worth. The smallest diamonds are sorted into one category. Large diamonds, bigger than fifteen carats, are sorted into another category.

Gem vs. Industrial

Most mines produce many more industrial diamonds than gems. Here are the averages for a mine in Australia:

- 1% pink
- 27% near colorless to light yellow
- 72% industrial

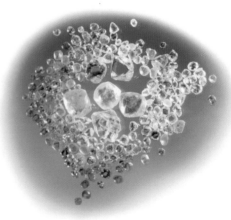

▲ Most diamonds are white to light yellow. The clearest diamonds, called whites, are considered the best stones.

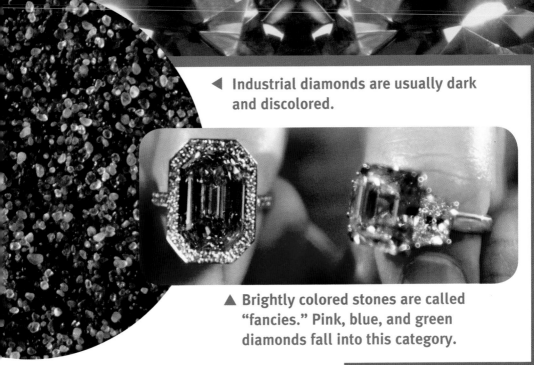

◀ Industrial diamonds are usually dark and discolored.

▲ Brightly colored stones are called "fancies." Pink, blue, and green diamonds fall into this category.

Diamonds are also sorted by color and clarity. People who sell diamonds created two scales and rank diamonds by these scales.

One scale grades the diamonds based on their color. Diamonds that are the most colorless are considered the best. The other scale ranks the gems based on their clarity. Clarity is how transparent a stone is. The best stones are called flawless.

How much a diamond costs is determined by all of these things. A two-carat diamond is usually worth more than a one-carat diamond. But a one-carat white stone with no flaws would be worth more than a two-carat yellow stone with large flaws.

CAREERS
GEMOLOGIST

Are you fascinated by beautiful stones? Gemologists learn to identify and work with stones. Gemologists can work in many places. Many sell and repair jewelry, and some make jewelry. Gemologists can also work for insurance companies. They appraise, or decide the value of, different stones. To become a gemologist, you have to finish a six-month program that teaches you about gems.

▲ Today, many diamonds are cut in India.

Cutting Diamonds

Before they are cut, diamonds are called "roughs." They don't look anything like the diamonds you see in jewelry. To make a diamond sparkle, facets, or flat planes, are cut into it. The facets catch and reflect light.

Cutting a diamond is tricky. A diamond cutter will study a stone carefully first. He needs to find the stone's cleavage. The cleavage is where the diamond will cut most evenly. If the cutter cuts the stone the wrong way, he can destroy it. He also wants to find any flaws in the stone. Flaws can sometimes be cut out of the stone. A talented diamond cutter can also hide flaws when he cuts a stone.

▼ Diamond cutters have experimented for years to find the best way to cut a diamond. Today, many use a design called the brilliant cut. In this design, fifty-eight facets are cut into the diamond.

Diamond cutters try to cut away as little of the stone as possible. The larger the stone, the higher the price.

When the cutter is ready, he will mark the cut lines on the diamond. The next step is called cleaving. That is when he cuts the diamond in two. Then he will begin to cut the facets into the stone. Finally, the stone is polished.

▲ Because only a diamond will cut another diamond, diamond cutters coat their tools with diamond dust.

IT'S A FACT

In 1988, miners in South Africa discovered a 599-carat stone. It was named the Centenary Diamond. Diamond cutters studied the stone for a year before they started to cut it. They took another year to do the work. The finished stone weighs 273 carats and has 247 facets.

Diamonds at Work

Most diamonds don't end up as necklaces or rings. Instead, they are used for tools. Industrial diamonds aren't pretty, but they still have the same properties as gem-quality diamonds. They are the hardest material available. They are also durable and can withstand heat.

Many of the drills that are used to dig oil wells or bore tunnels through rock have diamond tips. The rovers that were sent to explore the surface of Mars were equipped with diamond-tipped drills. NASA engineers knew the drills would be strong enough to collect rock samples.

▼ A diamond-tipped drill may cost more money, but it will last a long time.

IT'S A FACT

Most industrial diamonds come from mines in Australia.

Diamonds can also be used for more delicate cutting. Thin slices of diamonds are used to make scalpels. Diamond blades are used to cut the lenses for eyeglasses. The drill a dentist uses also has a diamond tip.

Diamonds can be used for windows, too. Because diamonds are transparent and durable, they were used to make the windows for spacecraft. A diamond coating can also be applied to windows. The coating protects the windows from scratches.

Diamond grit, or very small pieces of diamonds, is another form of industrial diamond. Diamond grit is often used to polish things.

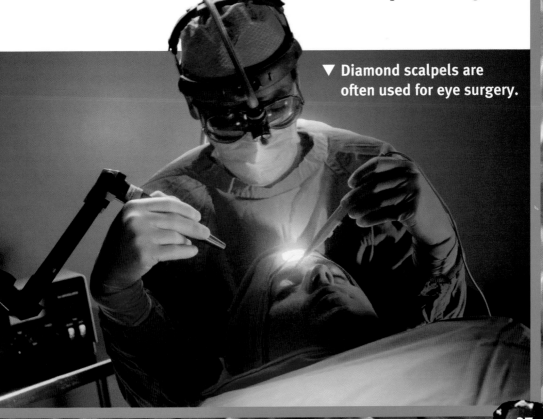

▼ Diamond scalpels are often used for eye surgery.

Growing Diamonds

Diamonds are valuable. So it's not surprising that people wanted to find a way to make diamonds. As early as 1880, people started to try. No one made much progress.

In 1950, a group of scientists at the General Electric Company began to work on the problem. They built a machine called a diamond press that could generate heat and pressure. The scientists experimented with graphite, which, like diamond, is made of carbon.

After five years of work, the scientists found a way to turn graphite into diamond. The first **synthetic** (sin-THEH-tik) or artificial, diamonds were very tiny—about the size of a grain of sand—but they were still diamonds.

▲ Dr. Herbert Strong (right) is one of the scientists who created the first synthetic diamonds.

Since then, scientists have become better at making diamonds. They have built new machines that can make bigger stones. Some stones are as big as three carats. It takes about three days for a machine to make a diamond that big.

Today, many industrial diamonds are synthetic. People are finding new ways to use them, too. One of the most exciting ideas is to use synthetic diamonds in computers. Today, the microchips inside computers are made of silicon. Computers generate heat, and silicon holds up well. But as computers become more powerful, they generate more heat. If there is too much heat, silicon melts. Diamonds are able to withstand much hotter temperatures. That is why some people think computer chips should be made of diamonds.

▲ Early diamond machines did not run for very long without breaking. Today's machines can run for days at a time.

MATH MATTERS

What is the difference between real diamonds and synthetic diamonds? One difference is cost. Industrial diamonds cost between $7.00 and $10.00 per carat. Synthetic diamonds cost between $1.50 and $3.00 per carat.

Conclusion

People first found diamonds more than 2,000 years ago. Over the years, people have learned a lot about diamonds. They have learned where diamonds formed, how they formed, and how to find them. They have produced synthetic diamonds.

Yet there are still more things to discover. Where will the next diamonds be found? What new technology will help geologists find them? What new uses will people find for diamonds? Will there someday be diamonds inside everyone's computer? These are all questions that will have to be answered in the future.

Great Moments in Diamond History

4th century B.C.	Diamonds are discovered in India.
1600s	Explorer Jean-Baptiste Tavernier writes about diamonds and brings them back to Europe.
1700s	Diamonds are discovered in Brazil.
1867	Diamonds are discovered in South Africa.
1871	Underground source of diamonds is discovered in South Africa.
1955	First synthetic diamond is produced.
1961	Diamonds are mined from the ocean.
1973	Geologist John Gurney discovers G-10 garnet.

Glossary

atom
(A-tum) the smallest particle of an element that has the characteristics of that element (page 7)

carat
(KAIR-ut) a unit of weight for diamonds and other gems (page 14)

crust
(KRUST) Earth's rocky outer layer (page 8)

element
(EH-leh-ment) a pure substance, such as oxygen, carbon, or iron, that cannot be broken down into simpler substances (page 7)

erode
(ih-RODE) to wear away by wind and water (page 11)

kimberlite pipe
(KIM-ber-lite PIPE) a volcanic pipe that carries diamonds from Earth's mantle to the surface (page 8)

luster
(LUS-ter) the way a mineral reflects light (page 5)

mantle
(MAN-tul) the layer of rocks between Earth's crust and core (page 8)

mineral
(MIH-nuh-rul) a solid naturally occurring substance that forms in crystals (page 4)

primary deposit
(PRY-mair-ee dih-PAH-zit) mineral found in the place where it formed (page 11)

property
(PRAH-per-tee) a quality that can be measured or observed, such as hardness or flexibility (page 3)

refract
(rih-FRAKT) to bend (page 6)

secondary deposit
(SEH-kun-dair-ee dih-PAH-zit) mineral found away from where it was formed (page 11)

synthetic
(sin-THEH-tik) man-made (page 28)

Index